A Wolf's World

by Sara E. Turner

MW01011969

Contents

Science Vocabulary

living
Living things are alive.

Wolves are **living** things.

nonliving

Nonliving things are not alive and never were.

rocks

Rocks are **nonliving** things.

basic needs

Basic needs are what living things must have to stay alive.

Water is only one of a wolf's **basic needs.**

shelter

A **shelter** is a safe place where a living thing can make its home and grow.

| basic needs |
| living |
| nonliving |
| nutrients |
| shelter |

A hollow in the ground, or a den, is a wolf's **shelter.**

nutrients

Nutrients are parts of food and soil.

Wolf pups get **nutrients** from milk.

7

Living and Nonliving Things

These wolves rest near some rocks.
Wolves are **living** things.

living
Living things are alive.

Rocks are **nonliving** things.

nonliving

Nonliving things are not alive and
never were.

You can compare a wolf to a rock.

wolf

rock

Living Thing: Wolf	Nonliving Thing: Rock
• A wolf must have air.	• A rock doesn't need air.
• A wolf must have food and water.	• A rock doesn't need food and water.
• A wolf moves on its own.	• A rock can't move on its own.
• A wolf can grow.	• A rock can't grow.

Basic Needs of Animals

Living things have **basic needs.**

One basic need is water.

This wolf must have water to stay alive.

basic needs

Basic needs are what living things must have to stay alive.

Nonliving things do not have basic needs.

These rocks do not need water.

Living things must have air.

These wolf pups breathe in air.

Living things must have **shelter.**

A den is a shelter for wolves.

This den is an opening in the ground.

shelter

A **shelter** is a safe place where a living thing can make its home and grow.

Living things must have food.

A young pup drinks its mother's milk.
The pup gets **nutrients** from the milk.

nutrients

Nutrients are parts of food and soil.

Older wolves eat meat.

They hunt animals.

A wolf hunts small animals without its
pack, or group.

The wolf shares the meat with the pups.

Basic Needs of Plants

Many wolves live in a forest.

Plants live there, too.

Like all living things, plants have basic needs.

Fir trees must have air, water, and light.

Fir trees grow where wolves live.

The parts of a fir tree work together to keep it alive.

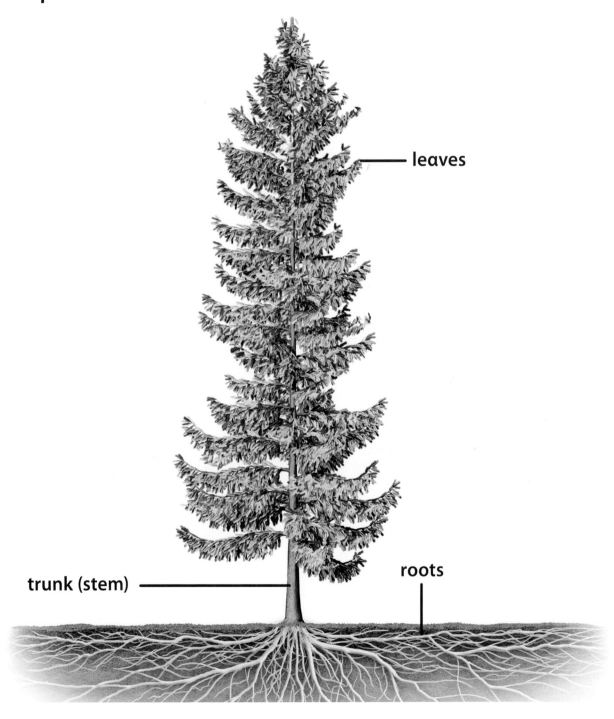

leaves

trunk (stem)

roots

Plants must have water, too.

These trees have plenty of water.

When it doesn't rain, plants don't get the water they need.

These trees need more water to grow.

Plants in a forest must have space to grow.

The forest has lots of space for plants and wolves to live and grow.

Conclusion

Wolves and fir trees are living things in a forest. There are nonliving things in a forest, too. The living things have basic needs. The nonliving things do not.

Think About the Big Ideas

1. How are living and nonliving things different?
2. What are the basic needs of fir trees?
3. What are the basic needs of wolves?

Share and Compare

Turn and Talk

Compare the living and nonliving things in your books. How are they different?

Read

Find a photo with a caption and read it to a classmate.

Write

Describe the basic needs of a living thing in your book. Share what you wrote with a classmate.

Draw

Show an animal getting something it must have to stay alive. Share your drawing with a classmate.

Meet Mireya Mayor

Scientists work together to solve problems. They try many ways to answer their questions.

Mireya Mayor helped discover a mouse lemur. She and her team wanted to catch it. They used a water bottle to make a trap, but the trap didn't work. The team finally found a way to catch the lemur. How do you think they did it?

Mireya Mayor holds a mouse lemur.

Index

Acknowledgments
Grateful acknowledgment is given to the authors, artists, photographers, museums, publishers, and agents for permission to reprint copyrighted material. Every effort has been made to secure the appropriate permission. If any omissions have been made or if corrections are required, please contact the Publisher.

Photographic Credits
Cover (bg) Ronnie Howard/Shutterstock; Cvr Flap (t) John Pitcher/iStockphoto, (c) Elena Elisseeva/Shutterstock, (b) Digistock; Title (bg) John Pitcher/iStockphoto; 2-3 Alan D. Carey/PhotoDisc/Getty Images; 4, 5, 8-9, 11 (r) John Pitcher/iStockphoto; 6, 12 Digistock; 7 (t), 15 Robert Winslow/Animals Animals; 7 (b), 17 Thomas & Pat Leeson/Photo Researchers, Inc.; 10-11 (bg), 28 Digistock; 11 (l), 27 (inset) Digistock; 13 Tim Laman/National Geographic Image Collection; 14 John Pitcher/iStockphoto; 16 FranBurek/Corbis Premium RF/Alamy Images; 18 Creatas/Jupiterimages; 19 Arco Images GmbH/Alamy Images; 20-21 Digistock; 22 Sergey Dubrovskiy/iStockphoto; 24 Elena Elisseeva/Shutterstock; 25 Aistov Alexey/Shutterstock; 26-27 (bg) Mark Ahn/Shutterstock; 30-31 Mark Thiessen, National Geographic Photographer; Inside Back Cover (bg) Geoff Kuchera/iStockphoto.

Illustrator Credits
23 Paul Mirocha

Neither the Publisher nor the authors shall be liable for any damage that may be caused or sustained or result from conducting any of the activities in this publication without specifically following instructions, undertaking the activities without proper supervision, or failing to comply with the cautions contained herein.

Program Authors
Randy Bell, Ph.D., Associate Professor of Science Education, University of Virginia, Charlottesville, Virginia; Malcolm B. Butler, Ph.D., Associate Professor of Science Education, University of South Florida, St. Petersburg, Florida; Kathy Cabe Trundle, Ph.D., Associate Professor of Early Childhood Science Education, The Ohio State University, Columbus, Ohio; Nell K. Duke, Ed.D., Co-Director of the Literacy Achievement Research Center and Professor of Teacher Education and Educational Psychology, Michigan State University, East Lansing, Michigan; Judith Sweeney Lederman, Ph.D., Director of Teacher Education and Associate Professor of Science Education, Department of Mathematics and Science Education, Illinois Institute of Technology, Chicago, Illinois; David W. Moore, Ph.D., Professor of Education, College of Teacher Education and Leadership, Arizona State University, Tempe, Arizona

The National Geographic Society
John M. Fahey, Jr., President & Chief Executive Officer
Gilbert M. Grosvenor, Chairman of the Board

Copyright © 2011 The Hampton-Brown Company, Inc., a wholly owned subsidiary of the National Geographic Society, publishing under the imprints National Geographic School Publishing and Hampton-Brown.

All rights reserved. No part of this book may be reproduced or transmitted in any form or by any means, electronic or mechanical, including photocopying, recording, or by an information storage and retrieval system, without permission in writing from the Publisher.

National Geographic and the Yellow Border are registered trademarks of the National Geographic Society.

National Geographic School Publishing
Hampton-Brown
www.NGSP.com

Printed in the USA.
RR Donnelley, Wetmore, TX

ISBN: 978-0-7362-7576-7

11 12 13 14 15 16 17

10 9 8 7 6 5 4 3